LUNAR LANDER QUEST:

Humanity's Journey To Touch The Moon

Charlie S. Smith

TABLE OF CONTENT

INTRODUCTION

In the vast universe, there exists a celestial body that has captured the human imagination for centuries, our closest cosmic neighbor, the Moon. A radiant orb suspended in the night sky, the Moon has inspired myths, sparked dreams and, most notably, been the focus of one of humanity's greatest endeavors.

Begins a historic odyssey, tracing the Moon's profound impact on our collective consciousness and quest extraordinary to bring humans to its desolate surface. From ancient myths and early lunar observations to the cutting-edge technology of the Apollo missions and the evolving landscape of modern lunar exploration, this book explores a journey never before seen that ends up touching the Moon and deciphering its mysteries.

 The story unfolds through the lens of science, technology, and human ingenuity, revealing the trials and triumphs of pioneers who dared to dream beyond Earth's borders. As we turn the pages of this exploration, we encounter the political context of the space race, the audacity of the Apollo missions, and the indomitable spirit that propelled humanity to achieve the impossible. Beyond the pages of history, we explore the complex design and engineering challenges of lunar landers, the scientific discoveries that have reshaped our understanding of the Moon and Earth land as well as global cooperation that has overcome political divisions in the quest for knowledge.

The story extends to contemporary initiatives, such as the participation of private companies in the space race and the Artemis program's ambitious goal of returning humans to the surface moon. Lunar Lander Quest is not simply a chronicle of technological achievements, it is a celebration of the unwavering determination of the human spirit to reach new frontiers. The cultural impact of lunar exploration, the enduring legacy of the Apollo missions, and the transformative power of space exploration on our understanding of ourselves and our place in the universe. Across the pages, we embark on a celestial journey, following the trail left by brave explorers who ventured beyond our world to touch the Moon.

CHAPTER ONE: WHAT IS A LUNAR LANDER?

A lunar lander is a spacecraft designed to land on the face of the Moon. As of 2023, the Apollo Lunar Module is the only lunar lander to have ever been used in mortal spaceflight, completing six lunar levees from 1969 to 1972 during the United States' Apollo Program. Several robotic landers have reached the face, and some have returned samples to Earth. The design conditions for these landers depend on factors assessed by the cargo, flight rate, propulsive conditions, and configuration constraints. Other important design factors include overall energy conditions, charge duration, the type of charge operations on the lunar face, and the life support system if crewed. The fairly high gravity(advanced than all known asteroids, but lower than all Solar System globes) and lack of a lunar atmosphere negate the use of aerobraking, so a lander must use propulsion to decelerate and achieve a soft wharf.

In a noteworthy lunar achievement, the primary private spacecraft to arrive effectively on the Moon touched down on 22 February. The spacecraft, named Odysseus and built by Instinctive Machines in Houston, Texas, moreover got to be the primary US lunar lander in 1972, when the final team of Apollo space travelers went by the Moon. Odysseus advertised up a few nail-biting minutes within the hours sometime recently of landing, such as the glitch of the laser rangefinders that were assumed to assist direct its independent travel down to the lunar surface. Mission engineers had to transfer a software fix to jury-rig it to utilize an auxiliary laser given by NASA instep.

The precise state of the spacecraft remained hazy promptly after its landing, which happened at 5:23 p.m. Houston time. But it was sending a swoon flag back to mission control in Houston, showing that at least a few parcels of it had survived the landing.

"Odysseus has found its modern home," said mission executive Tim Crain as the control room burst into cheers. After a nail-biting plummet, the Odysseus spacecraft lands close to the lunar south post and plans to kick off a week of data-gathering. Notwithstanding how operational the spacecraft might be going forward, the landing is a major shot within the arm for US and commercial endeavors to return to the Moon. NASA paid for much of the private mission and is checking on companies such as Instinctive Machines to assist ship hardware and logically rebellious to the Moon in preparation for returning space travelers there. "The US has returned to the Moon," said NASA director Nelson.

1.1: History Of Lunar From 1958-2024

The Luna program was a series of robotic impactors, flybys, orbiters, and landers flown by the Soviet Union between 1958 and 1976. Luna 9 was the first spacecraft to achieve a soft wharf on the Moon on February 3, 1966, after 11 unprofitable attempts. Three Luna Spacecraft returned lunar soil samples to Earth from 1972 to 1976. Two other Luna spacecraft soft-landed the Lunokhod robotic lunar rover in 1970 and 1973. Luna achieved an aggregate of seven successful soft levees out of 27 wharf attempts. The United States Surveyor program first soft-landed Surveyor 1 on June 2, 1966, this original success was followed by four fresh successful soft-levees, the last being on January 10, 1968. The Surveyor program achieved an aggregate of five successful soft levees out of seven wharf attempts through January 10, 1968. As of 2024, Apollo Lunar is the only crewed lunar lander.

The Apollo program completed six successful lunar soft levees from 1969 until 1972; a seventh lunar wharf attempt by the Apollo program was abandoned when Apollo 13's service module suffered explosive venting from its oxygen tanks. Several LK lunar modules were flown without crew in the low Earth route, but the LK lunar module norway flew to the Moon, as the development of the N1 Rocket Launch Vehicle needed for the lunar flight suffered lapses(including several launch failures), and after the first mortal Moon levees were achieved by the United States, the Soviet Union canceled both the N1 Rocket and the LK Lunar Module programs without any further development.

The Chinese Lunar Exploration Program(also known as the theChang'e design) includes a robotic lander, rover, and sample-return factors; the program realized an original successful lunar soft-wharf with theChang'e 3 spacecraft on 14 December 2013. As of 2023, the CLEP has achieved three successful soft-levees out of three wharf attempts, videlicetChang'e 3, Chang'e 4, and Chang'e 5. Chang'e 4 made history by making humanity's first-ever soft wharf on the far side of the moon. Israel's SpaceIL tried a robotic lunar wharf with its Beresheet lander on 4 April 2019; the attempt failed.

As of 2023, SpaceIL has plans for another soft-wharf attempt using a follow-up robotic lander named Beresheet 2. India's Chandrayaan Programme conducted an unprofitable robotic lunar soft-wharf attempt on 6 September 2019 as part of its Chandrayaan- 2 spacecraft with the lander crashing on the Moon's face. On 23 August 2023, the program's follow-up Chandrayaan- 3 landers achieved India's first robotic soft wharf.

 Japan's space agency (not to be confused with China's i- I-Space) tried a lunar soft wharf by its Hakuto- R Mission 1 robotic lander on 25 April 2023. The attempt was unprofitable and the lander crashed into the lunar surface. The company presently has plans for another wharf attempt in 2024. Russia's LunaGlob program, the successor program to the Soviet Union's Luna program, launched the Luna 25 lunar lander on 10 August 2023; the inquiry's intended destination was near the lunar south pole, but on 19 August 2023 the lander crashed on the Moon's face.

Japan's Smart Lander for probing the Moon launched on 6 September 2023. The inquiry made a successful lunar wharf on 19 January 2024, near the

Shioli crater on the lunar nearside; the lander carried two small rovers on board. It landed on 19 January 2024 at 1520 UTC, making Japan the 5th country to soft land on the moon. Although it landed successfully, it's in the wrong station because the solar panels are acquainted westwards facing opposite the Sun at the launch of the lunar day, thereby failing to induce enough power. The lander operated on internal battery power, which was completely drained that day. The charge's drivers hope that the lander will wake up after many days when the sun should hit the solar panels.

Irrespective of this solar array issue on the lander, the two LEV 1 and 2 rovers, stationed during swimming just before the final wharf are working as anticipated, and LEV-1 communicating singly to the ground stations. LEV-1 conducted six hops on the lunar face. Images taken by LEV-2 show the wrong station wharf with the loss of a machine snoot during descent and indeed possible sustained damage to the lander's Earth-bound antenna, which isn't refocused towards Earth. Irrespective of the wrong station and loss of communication with the lander, the charge was formally successful after evidence that its primary thing of wharf within 100 m(330 ft) of its wharf spot was formally achieved.

On 8 January 2024, the first charge of the NASA-funded CLPS program, Peregrine Mission One, was placed into an elliptical High Earth route; the charge's lunar lander was anticipated to execute several propulsive pushes to achieve a low lunar route before making a lunar wharf attempt on 23 February 2024. Still, an energy leak was detected on the spacecraft several hours after launch with the result that it would lose its capability to maintain station control and to charge its battery, thereby rendering the accession of the lunar route doubtful and preventing a wharf attempt. IM- 1 Nova- C

Odysseus launched on 15 February 2024 towards the Moon via Falcon 9 on a direct intercept line. On 22 February 2024, Intuitive Machine's Odysseus successfully landed on the Moon after taking off on a SpaceX Falcon 9 takeoff on 15 February 2024 in a charge between NASA, SpaceX, and Intuitive Machines, marking the United States' first soft unmanned moon wharf in over 50 times. This charge also marks the first intimately possessed spacecraft to land on the Moon and the first wharf with cryogenic forces.

CHAPTER TWO: EARLY LUNAR DREAMS

It refers to the historical and cultural fascination that humans have had with the Moon for centuries. These dreams are often reflected in mythology, literature, and early scientific speculations.

Early moon dreams reflect the human desire to understand, interpret, and explore the celestial bodies that share our night sky.

These dreams laid the foundation for subsequent scientific discoveries, eventually leading to human travel to the Moon and beyond. In many ancient cultures, the Moon is personified in mythology and folklore. Deities associated with the Moon, such as Selene in Greek mythology or Chang'e in Chinese mythology, illustrate their cultural importance and early attempts to explain lunar phenomena. Before the invention of telescopes, people observed the Moon with the naked eye. They noticed its changing phases and frequent eclipses, sparking curiosity and prompting different cultures to create stories and explanations for these celestial events.

It is a frequent theme in literature, poetry, and philosophical works. Writers such as Shakespeare, who often referred to the Moon in their works, contributed to romantic and symbolic associations with the Earth's satellite. Early astronomers and philosophers, such as Galileo Galilei, turned their telescopes toward the Moon, making detailed observations and challenging existing beliefs. Galileo's observations, including of mountains and craters on the Moon, provided experimental evidence that the Moon was not a perfect, flat sphere.

The remote and seemingly inaccessible nature of the Moon has sparked the imagination about what might exist on its surface. Before the age of space exploration, the Moon was often considered a mysterious kingdom, inspiring stories of hidden civilizations or fantastical landscapes. Even before the possibility of space travel, writers such as Jules Verne and H.G. Wells explored the concept of lunar travel in their science fiction works.

These early literary works laid the foundation for future discussions of space exploration. It has influenced many different aspects of human culture, from calendar systems based on lunar phases to the Moon's role in art, music, and religious practice. The moon is a powerful symbol representing cycles, femininity, mystery, and enlightenment. Its symbolism is found in ancient symbols, flags, and coats of arms of different civilizations.

CHAPTER THREE: THE SPACE RACE

The Space Race was a 20th-century competition between two Cold War rivals, the United States and the Soviet Union, to realize predominant spaceflight capability. It had its beginnings within the ballistic missile-based atomic arms race between the two countries after World War II and had its crest with the more specific Moon Race to arrive on the Moon between the US moonshot and Soviet moonshot programs.

The mechanical advantage illustrated by spaceflight accomplishment was seen as essential for national security and became a portion of the imagery and philosophy of the time. The Space Race brought spearheading dispatches of counterfeit satellites, automated space tests to the Moon, Venus, and

Mars, and human spaceflight in an Earth circle and eventually to the Moon.

Open intrigue in space travel began with the 1951 distribution of a Soviet youth magazine and was instantly picked up by US magazines. The competition started on July 30, 1955, when the United States reported its aim to dispatch fake satellites for the Universal Geophysical Year. Four days later, the Soviet Union reacted by pronouncing they would too dispatch an adherent "within the close future". The propelling of satellites was empowered by advancements in ballistic rocket capabilities since the conclusion of World War II.

The competition picked up Western open consideration with the "Sputnik emergency" when the USSR achieved the beginning with effective today's dispatch, Sputnik 1, on October 4, 1957. It picked up force when the USSR sent the primary human, Yuri Gagarin, into space with the orbital flight of Vostok 1 on April 12, 1961. These were followed by a string of other early firsts accomplished by the Soviets over the following long time.

Gagarin's flight drove US President John F. Kennedy to raise the stakes on May 25, 1961, by asking the US Congress to commit to the objective of "landing a man on the Moon and returning him securely to the Earth" sometime recently the conclusion of the decade. Both nations began developing super heavy-lift dispatch vehicles, with the US effectively conveying the Saturn V, which was huge and sufficient to send a three-person orbiter and two-person lander to the Moon.

Kennedy's Moon landing objective was accomplished in July 1969, with the flight of Apollo 11. The conclusion of Apollo 11 is respected by numerous Americans as ending the Space Race with an American triumph. This view is challenged by a few history specialists, while space history specialist Asif A. Siddiqi proposed a more adjusted view. The USSR proceeded to seek to run lunar programs but did not succeed, with its N1 rocket to dispatch and land on the Moon sometime recently the US inevitably canceled it to concentrate on Salyut, the first space station program, and the primary arrivals on Venus and Defaces. In the interim, the US landed five more Apollo groups on the Moon and proceeded to investigate other extraterrestrial bodies robotically.

A period of détente taken after the April 1972 agreement on an agreeable Apollo–Soyuz Test Venture (ASTP), coming about within the July 1975 meeting in the Earth circle of a US space traveler team with a Soviet cosmonaut crew and joint improvement of a universal docking standard

APAS-75. Being considered as the ultimate act of the Space Race, the competition was as it were continuously supplanted with participation. The collapse of the Soviet Union, in the long run, permitted the US and the recently established Russian League to conclude their Cold War competition too in space, by concurring in 1993 on the Shuttle–Mir and Worldwide Space Station programs.

CHAPTER FOUR: FIRST MANUFACTURED SATELLITES

In 1955, with both the United States and the Soviet Union building ballistic rockets that might be utilized to dispatch objects into space, the arrangement was set for nationalistic competition. On July 29, 1955, James C. Hagerty, President Dwight D. Eisenhower's press secretary reported that the United States aimed to dispatch "little Earth circling satellites" between July 1, 1957, and December 31, 1958, as a portion of the US commitment to the Universal Geophysical Year (IGY). On Eminent 2, at the 6th Congress of the Worldwide Astronautical League in Copenhagen, researcher Leonid I.

Sedov told universal correspondents at the Soviet international haven of his country's intention to dispatch a partisan as well, within the "close future" Sputnik, an arrangement of three manufactured Earth satellites, the primary of whose launch by the Soviet Union on October 4, 1957, introduced the space age. Sputnik 1, the primary manufactured satellite propelled, was an 83.6-kg (184-pound) capsule.

The dispatch of Sputnik 1 stunned numerous Americans, who had accepted that their nation was mechanically ahead of the Soviet Union, and driven to the "space race" between the two nations. Sputnik 2, propelled on November 3, 1957, carried the puppy Laika, the primary living animal to be shot into space and circle Earth. Laika was a stray pooch found on the lanes of Moscow.

There were no plans to return her to Earth, and she lived only many hours in a circle. Sputnik 3, propelled on May 15, 1958, carried 12 rebellious to think about Earth's upper environment and space and was moreover the heaviest satellite to that time, weighing 1,327 kg (2,926 pounds). Sputnik 3 was initially aimed to be the primary disciple, but its complexity and measure drove the Soviets to dispatch the much less complex Sputnik 1 to defeat the United States into space.

The Soviets authoritatively called it three satellites Sputnik. Within the West, be that as it may, Sputnik was utilized as a nonexclusive title for Soviet satellites. These "Sputniks" included the primary tests to Venus (Venera 1) and Mars (Mars 1), as well as five missions within the Korabl-Sputnik program, which tried the manned Vostok shuttle sometime recently Yuri Gagarin's flight in 1961.

4.1: Project Apollo

Apollo, the project conducted by the U.S. National Flight and Space Organization (NASA) in the 1960s and '70s landed the primary people on the Moon. All told, 24 Apollo space travelers went by the Moon and 12 of them strolled on its surface. Extra NASA space explorers are planned to return to the Moon by 2025 as a portion of the Artemis space program. In May 1961 Pres. John F. Kennedy committed America to landing space travelers on the Moon by 1970. The choice among competing procedures for accomplishing a Moon landing and return was not settled until impressive encouragement pondered.

Three strategies were considered. In coordinate rising, one vehicle would lift off from Earth, arrive on the Moon, and return. Be that as it may, the proposed Nova rocket would not be prepared by

1970. In Earth circle meet, a shuttle carrying the group would dock with the impetus unit that would carry sufficient fuel to go to the Moon. However, this strategy required two partitioned dispatches. Within the strategy eventually utilized, lunar circle meet, a capable dispatch vehicle (Saturn V rocket) put a 50-ton shuttle in a lunar direction. The shuttle had three parts. The funnel-shaped command module (CM) carried three space explorers.

One space explorer remained within the CSM while the other two landed on the Moon within the LM. The LM had a plunge arranged and a climb organized. The plummet stage was cleared out on the Moon, and the space travelers returned to the CSM within the rising organism, which was disposed of in the lunar circle. The LM was flown as if it were within the vacuum of space, so streamlined contemplations did not influence its plan. (Hence, the LM has been called the primary "true" shuttle.)

Sometime recently reentering Earth's environment, the SM was discarded to burn up.

The CM sprinkled down within the sea. The lunar circle meet had the points of interest of requiring as it were one rocket and of sparing fuel and mass since the LM did not require to return to Earth. Uncrewed missions testing Apollo and the Saturn rocket started in February 1966. The primary run Apollo flight was deferred by an awful mishap, a fire that broke out within the Apollo 1 shuttle amid a ground practice on January 27, 1967, murdering space explorers Virgil Grissom, Edward White, and Roger Chaffee. NASA reacted by delaying the program to create changes such as not employing pure oxygen air at dispatch and supplanting the CM hatch with one that might be opened rapidly.

In October 1968, after several uncrewed Earth-orbit flights, Apollo 7 made a 163-orbit flight carrying a full group of three space explorers. Apollo 8 carried out the primary step of running a lunar investigation. Earth's circle was injected into a lunar direction, completed the lunar circle, and returned securely to Earth. Apollo 9 carried out a drawn-out mission in Earth Circle to check out the LM. Apollo 10 traveled to a lunar circle and tried the LM to inside 15.2 km (9.4 miles) of the Moon's surface. Apollo 11, in July 1969, climaxed the step-by-step strategy with a lunar landing; on July 20 space explorers Neil Armstrong and Edwin ("Buzz") Aldrin were the first people to set foot on the Moon's surface.

Apollo 13, propelled in April 1970, endured a mishap caused by a blast in an oxygen tank but returned securely to Earth as the sun-powered wind tried and the seismographic estimations of the lunar surface. Starting with Apollo 15, space explorers drove a lunar wanderer to the Moon. Apollo 17, the ultimate flight of the program, took off in December 1972.

In October 1968, after several uncrewed Earth orbit flights, Apollo 7 made a 163-orbit flight carrying a full group of three space explorers. Apollo 8 carried out the primary step of running a lunar investigation. Earth's circle was injected into a lunar direction, completed the lunar circle, and returned securely to Earth. Apollo 9 carried out a drawn-out mission in Earth Circle to check out the LM. Apollo 10 traveled to a lunar circle and tried the LM to inside 15.2 km (9.4 miles) of the Moon's surface. Apollo 11, in July 1969, climaxed the step-by-step strategy with a lunar landing; on July 20 space explorers Neil Armstrong and Edwin ("Buzz") Aldrin was the first person to set foot on the Moon's surface.

Apollo 13, propelled in April 1970, endured a

mishap caused by a blast in an oxygen tank but returned securely to Earth as the sun-powered wind tried and the seismographic estimations of the lunar surface. Starting with Apollo 15, Space explorers drove a lunar wanderer to the Moon. Apollo 17, the ultimate flight of the program, took off in December 1972.

4.2: Apollo 11 First Steps On The Moon

Apollo 11 First Steps on the Moon Apollo 11 was a historic space mission led by NASA, the United States space agency, and it became the first manned mission to successfully land on the Moon.

The primary goal of the mission was to achieve President John F. Kennedy's goal of sending an astronaut to the Moon and returning him safely to Earth before the end of the 1960s. The Apollo 11 mission achieved this feat. This epic achievement and the first steps on the Moon were photographed by astronauts Neil Armstrong and Buzz Aldrin.

Apollo 11 was launched on July 16, 1969, on a Saturn V rocket from the Kennedy Space Center in Florida. The spacecraft included the command module "Columbia", where Michael Collins orbited the Moon, and the lunar module "Eagle", which carried Armstrong and Aldrin to the lunar surface.

After reaching the lunar orbit, Armstrong and Aldrin entered the lunar module and separated from the command module. On July 20, 1969, the lunar module began its descent to the Moon's surface. The lunar module landed on the Moon's surface in an area known as the Sea of Tranquility. Landing was a critical phase and Armstrong manually steered the module to a safe landing point with only a few seconds of fuel remaining. Neil Armstrong's famous words, "The Eagle has landed", were transmitted to Mission Control and marked the successful landing of the lunar module on the surface of the Moon at 20: 17 UTC on July 20, 1969.

First not on the Moon (July 20, 1969) About six hours after landing, Neil Armstrong descended the lunar module ladder and set foot on the Moon, uttering the historic words: "That's one small step for a man, one giant leap for mankind." Buzz Aldrin joined him soon after. Armstrong and Aldrin conducted various scientific experiments, deployed instruments, and collected lunar samples during their time on the Moon.

They also planted an American flag and left a sign reading: "Here come the people from the planet Earth set foot on the Moon for the first time after spending approximately 21 hours on the lunar surface, Armstrong and Aldrin returned to the lunar module." The ascent stage rendezvoused with the command module, where Michael Collins was orbiting the Moon. The astronauts returned safely to Earth and landed in the Pacific Ocean on July 24, 1969.

 The crew was picked up by the USS Hornet and spent several weeks in quarantine to guard against possible pathogens on the moon. The success of Apollo 11 was a decisive moment in human history, symbolizing the achievement of a seemingly impossible goal and demonstrating humanity's ability to explore space. The mission remains an enduring symbol of human ingenuity, determination, and the quest for knowledge.

CHAPTER FIVE: MAJOR TECHNOLOGICAL PROGRESS OF LUNAR LANDER LANDING

Future investigation requires major technological propels within the landing spacecraft. The mission employments demonstrated European innovations from past missions as much as conceivable. For this case, it incorporates engines altered from the Automated Transfer Vehicle. Landers will have to target well-illuminated landing locales and regions of special scientific intrigue and meet with other spacecraft. Landers will have to arrive delicately on the possibly unsafe landscapes so as not to damage sensitive gear, landing more cargo will be fundamental to extend mission returns and to begin building things on the target site.

Landers will get to work freely with people, making choices in genuine time. ESA has been creating Lunar Lander's direction, route, and control for a long time. In the same way that computer-driven cars may before long end up a reality on Earth, a European Lunar Lander may before long arrive on the Moon with high accuracy and no human intervention. The shuttle should know its position exceptionally precisely from the minute its mission begins. Though past missions were fulfilled with a landing precision of kilometers, the Lunar Lander will have a precision of some hundred meters. As there's no obsequious route on the Moon and depending on an Earth-based or inertial route isn't sufficient, these conventional procedures will be complemented with an image-based solution.

An optical route framework forms pictures of the lunar surface employing an effective computer on the spacecraft. The framework distinguishes points of interest such as cavities amid plummet and matches them with a set of points of interest put away in a database on the spacecraft. This will altogether move forward the lander's capacity to find its position. The database is made from data from missions such as NASA's Lunar Observation Orbiter and Japan's Kaguya. The route system will utilize the movement of the landscape watched by the camera to move forward with its assessment of the lander's speed. Making beyond any doubt these systems work accurately is troublesome since it requires near-perfect pictures of the Moon's surface that the lander will see amid its genuine flight.

Of course, numerous reenactments and tests are planned. Two minutes before landing, the landing location will come into Lunar Lander's field of sea and the computer's nitty-gritty investigation begins. At this arrangement, the route camera will be backed by a filtering lidar sensor, which employs laser beats to reconstruct landing location geography. Although the camera needs light to capture pictures of the Moon, the lidar is impenetrable to the harsh lighting conditions in the South Post region. Many questions had to be replied to by Lunar Lander's computer. Is the assigned landing location sunlit or within the shade? Are there huge boulders, cavities, or slants that seem to debilitate a secure landing? Is sufficient fuel left to alter course and reach a more secure location?

The spacecraft's brilliant systems will make choices and command the engines to control its direction. Twenty-seven thrusters, adjusted from other European missions, will work together to ensure that the Lunar Lander breaks from its orbital speed of 6000 km/h down to a few km/h while remaining on course to its landing location. Lunar Lander will be making all these decisions on its claim, more than

350,000 km from Earth. Once the landing arrangement is activated, mission control can as it were to sit back and observe, depending on Lunar Lander's innovation and its powerful computer to come to rest on the stark surface.

5.1: Testing and Prototypes

NASA initially conceived the Surveyor program in 1963 as a lander/orbiter combination venture but afterward scaled it down to soft landing. Each lander comprised a three-legged triangular aluminum structure with a huge strong fuel retro-rocket engine at the base. The lander was prepared with an advanced imaging framework. After three tests of the Atlas Centaur booster in 1965-1966, NASA propelled Surveyor 1 in May 1966. The mission was a reverberating victory. The spacecraft landed effectively within the Sea of Storms on June 2,

1966, and took more than 11,000 photographs of the surface over a month-long period.

Even though Surveyor 2 fizzled, Surveyor 3 effectively landed on the Moon in April 1967. In expansion to an imaging framework, the lander too included a remote scooper arm to decide the thickness of lunar soil. Tests appeared that the lunar soil had the consistency of damp sand. More than two a long time afterward, in November 1969, Apollo 12 space travelers Charles Conrad, Jr. and Alan Bean landed their Courageous Lunar Module around 180 meters from Surveyor 3 and recuperated some of its parts to assess the natural impacts of a long period on the Moon's surface.

Surveyor 4 was a disappointment, but Surveyors 5, 6, and 7 effectively landed on the Moon in 1967 and 1968, returning tremendous sums of photos and information on the Moon that were basic to planning tests for the Apollo missions. In addition, the five effective Surveyors returned more than 87,000 photographs of the Moon and appeared that it was attainable to soft-land an expansive test on the Earth's as it were characteristic adherent.

A model of one of the landing craft a Pittsburgh-based company is planning has been dispatched to NASA's Johnson Space Center in Houston for testing in the development of an arranged 2023 trip to the moon.``It's super-exciting. I never would have envisioned building a lunar lander in PaDaniel Gillies, mission executive for Ast robotic's Griffin Mission One. The Griffin model was dispatched to Houston before this month to test its plan. It is outlined to carry NASA's Volatiles Investigating Polar Exploration Rover (Viper) to the lunar south shaft, Gilles said.NASA will utilize the model to test the plan of Griffin's slopes to see on the off chance that the Viper can exit the lander to investigate the moon. The Griffin weighs more than 1,110 pounds and is the biggest landing make Astrobotic has outlined hence distant, Gilles said. In June, NASA granted Astrobotic a $200 million contract for Griffin. This can be the foremost progressed testing of the lander's plan, Gilles said A model of one of the landing plans a Pittsburgh-based company is planning has been dispatched to NASA's Johnson Space Center in Houston for testing in progress.

CHAPTER SIX: ARTEMIS PROGRAM RETURNING TO THE MOON

The Artemis program could be a Moon investigation program that's driven by the United States NASA and was formally set up in 2017 using Space Arrangement Order 1. The program's expressed long-term objective is to set up a lasting base on the Moon to encourage human missions to Mars. Two vital components of the Artemis program are determined from the now-cancelled Group of Stars program, the Orion shuttle, and the Space Dispatch Framework (as a resurrection of Ares V). Other components of the program, such as the Lunar Door space station and the Human Landing Framework, are under improvement by government space organizations and private spaceflight companies.

This collaboration is bound together by the Artemis Concurs and administrative contracts. The Space Dispatch Framework, Orion shuttle, and the Human Landing Framework frame the most spaceflight foundation for Artemis, and the Lunar Door plays a supporting part in human residence. Supporting frameworks for Artemis incorporate the Commercial Lunar Payload Administrations, Snake wanderer, advancement of ground frameworks, Artemis Base Camp on the Moon, Moon wanderers, and spacesuits.

On 30 June 2017, President Donald Trump marked an official arrangement to re-establish the National Space Committee, chaired by Vice President Mike Pence. The Trump administration's to begin with budget ask kept Obama-era human spaceflight programs in put. Commercial Resupply Services, Commercial Team Improvement, the Space Dispatch

Framework, and the Orion shuttle for profound space missions, while decreasing Earth science inquiries about and calling for the disposal of NASA's instruction office.

The Artemis program joins a few major components of already canceled NASA programs and missions, including the Group of Stars program and the Space Rock Divert Mission. Initially enacted by the NASA Authorization Act of 2005, Group of Stars included the improvement of Ares I, Ares V, and the Orion Group Investigation Vehicle. The program ran from the early 2000s until 2010.

 In May 2009, President Barack Obama set up the Augustine Committee to take under consideration several goals counting back to the Universal Space Station, advancement of missions past Moon Earth circle (counting the Moon, Mars, and near-Earth objects), and utilization of the commercial space industry inside characterized budget limits. The committee concluded that the Star grouping program was enormously underfunded and that a 2020 Moon landing was inconceivable. Star grouping along these lines is put on hold.

President Obama proposed US$6 billion in extra subsidizing and called for the advancement of a new heavy-lift rocket program to be prepared for development by 2015 with run missions to Mars circle by the mid-2030s. On 11 October 2010, President Obama marked into law the NASA Authorization Act of 2010, which included prerequisites for the prompt improvement of the Space Dispatch Framework as a follow-on dispatch vehicle to the Space Carry, and proceeded improvement of a Group Investigation Vehicle to be able of supporting missions past moo Soil circle beginning in 2016, whereas making utilize of the workforce, assets, and capabilities of the Space

Carry program, Star grouping program, and other NASA programs. The law also contributed to space innovations and mechanical autonomy capabilities tied to the general space investigation system, guaranteed proceeds back for Commercial Orbital Transportation Administrations and commercial Resupply Administrations, and extended the Commercial Group Advancement program.

On 11 December 2017, President Trump marked Space Policy Directive 1, an altar in the national space approach that gives for a U.S.-led, coordinated program with private segment accomplices for a human return to the Moon, taken after missions to Defaces and past. The arrangement calls for the NASA administrator to "lead an inventive and feasible program of investigation with commercial and international partners to empower human extension over the Sun-powered Framework and to bring back to Soil unused information and openings." The exertion is extraordinary to more successfully organize government, private industry, and universal endeavors toward returning people to the Moon and laying the foundation of inevitable human investigation of Mars.

The campaign (afterward named Artemis) draws upon bequest US shuttle programs, including the Orion space capsule, the Lunar Portal space station, and Commercial Lunar Payload Administrations, and creates entirely new programs such as the Human Landing Framework. The in-development Space Dispatch Framework is anticipated to serve as the essential dispatch vehicle for Orion, whereas commercial dispatch vehicles will dispatch different other components of the program.

On 26 March 2019, President Mike Pence reported that NASA's Moon landing objective would be quickened by four a long time with a planned landing in 2024. On 14 May 2019, NASA Chairman Jim Bridenstine declared that the modern program would be named Artemis, after the goddess of the Moon in Greek mythology who is the twin sister of Apollo. Despite the prompt unused objectives, deface missions by the 2030s were still expected as of May 2019. In mid-2019, NASA asked for US$1.6 billion in extra financing for Artemis for the monetary year 2020, whereas the Senate Apportionments Committee asked NASA for a five-year budget profile which is required for assessment and endorsement by Congress.

In February 2020, the White House asked for a subsidizing increment of 12% to cover the Artemis program as a portion of its financial year 2021 budget. The whole budget would have been US$25.2 billion per year with US$3.7 billion committed toward a Human Landing Framework. NASA Chief Financial Officer Jeff DeWit said he thought the agency includes "an exceptionally great shot" to push this budget through Congress despite Equitable concerns around the program. Be that as it may, in July 2020 the House Assignments Committee rejected the White House's request for subsidizing increment. The charge proposed within the House was devoted as it were US$700 million toward the Human Landing Framework, 81% (US$3 billion) short of the requested amount.

In April 2020, NASA granted subsidizing to Blue Root, Dynetics, and SpaceX for competing 10-month-long preparatory plans for the HLS. All through February 2021, Acting Chairman of NASA Steve Jurczyk repeated those budget concerns when inquiring about the project's plan, clarifying that "The 2024 lunar landing objective may not be a reasonable target. On 4 February 2021, the Biden organization supported the Artemis program. More particularly, White House Press Secretary Jen Psaki communicated the Biden administration's "bolster [for] this exertion and endeavor".

On 16 April 2021, NASA contract SpaceX to create, make, and fly two lunar landing flights with the Starship HLS lunar lander. Blue Root and Dynetics challenged the grant to the GAO on 26 April. After the GAO rejected the dissents, Blue Beginning sued NASA over the award, and NASA agreed to halt work on the contract until 1 November 2021 as the claim continued. The judge expelled the suit on 4 November 2021 and NASA continued work with SpaceX. On 25 September 2021, NASA released its, beginning with a computerized, intuitively realistic novel in celebration of National Comedian Book Day. "First Lady: NASA's Promise for Humankind" is the anecdotal story of Callie Rodriguez, the primary lady to investigate the Moon.

On 15 November 2021, a review of NASA's Office of Inspector Common evaluated the genuine fetch of the Artemis program at approximately $93 billion until 2025. In expansion to the initial SpaceX contract, NASA granted two rounds of isolated contracts in May 2019 and September 2021, on perspectives of the HLS to energize elective plans, independently from the introductory HLS advancement exertion. It declared in Walk 2022 that it was creating unused supportability rules and seeking both a Starship HLS overhaul (an alternative beneath the beginning SpaceX contract) and unused competing alternative designs. These came after feedback from individuals of Congress over the need for excess and competition and led NASA to inquire for extra back.

Some aspects of the program have been criticized, such as the utilization of near-rectilinear radiance circles and the maintainability of the space program. Orion's to begin with dispatch on the Space Dispatch Framework was initially set in 2016 but was rescheduled and launched on 16 November 2022 as the Artemis 1 mission, with robots and mannequins aboard. Agreeing to arrange, the crewed Artemis 2 dispatch will take out in late 2025, the Artemis 3 run lunar landing in 2026, the Artemis 4 docking with the Lunar Portal in 2028, and future annual arrivals on the Moon from there on.

The Artemis program is organized around a series of Space Dispatch Framework (SLS) missions. These space missions will increase in complexity and are planned to happen at intervals of a year or more. NASA and its accomplices have arranged Artemis 1 through Artemis 5 missions; afterward, Artemis missions have been proposed. Each SLS mission centers on the dispatch of an SLS dispatch vehicle carrying an Orion shuttle. Missions after Artemis 2 will depend on support missions propelled by other organizations and shuttle for back capacities.

Artemis 1 (2022) was the effective uncrewed test of the SLS and Orion and was the first test flight for both craft. The Artemis 1 mission set Orion into a lunar circle and after that returned to Soil. The SLS Piece 1 plan employs the ICPS second stage, which performs the trans-lunar infusion burn to send Orion to lunar space. For Artemis 1, Orion braked into a polar distant retrograde lunar circle and remained for almost six days before boosting back toward Soil.

The Orion capsule was isolated from its benefit module, re-entered the environment for aerobraking, and sprinkled down beneath parachutes. Artemis 1 was initially planned for late 2016, and as delays collected, in the long run for late 2021, but the dispatch date was pushed back to 29 Eminent 2022. Motor sensor issues caused a delay on that date; the following dispatch window was September 3.

A fuel supply line spill in a speedy detach arm on a ground tail benefit pole caused an encouraging delay for a period between 23 September and 4 October. Whereas the spill was mostly repaired to a satisfactory condition, climate delays due to Tropical Storm Ian constrained NASA directors to start preparing for the stack's rollback to the Vehicle Gathering Building and call off the September–early October dispatch window. In October 2022, NASA dispatch supervisors chose an unused launch date of

14 November, with reinforcement choices for 16 November and 19 November. In early November, NASA launch managers ruled out the 14 November choice and made arrangements to secure the SLS at the pad for Storm Nicole, after which launch was arranged for 16 November. On 16 November at 01:47:44 EST(06:47:44 UTC), Artemis 1 was effectively propelled from the Kennedy Space Center. Artemis 1 was completed at 09:40 PST (17:40 UTC) on 11 December, when the Orion shuttle sprinkles down in the Pacific Sea, west of Baja California, after a record-breaking mission, which saw Artemis travel more than 1.4 million miles on a way around the Moon sometime recently returning securely to Earth.

The splashdown happened 50 a long time to the day after NASA's Apollo 17 Moon landing, the final space traveler mission to touch down on the lunar surface. Implementation of the Artemis program will require extra programs, ventures, and commercial launchers to back the construction of the Gateway, launch resupply missions to the station, and convey various automated spacecraft and rebels to the lunar surface. A few precursor robotic missions are being facilitated through the Commercial Lunar Payload Services (CLPS) program, which is committed to scouting and characterization of lunar assets as well as testing principles for in-situ resource utilization.

Artemis 2 (2025) is planned to be the primary manned test flight of the SLS and the Orion shuttle. The four team individuals will perform broad testing in Earth circle, and Orion will at that point be boosted into a free-return trajectory around the Moon, which will return Orion to Soil for re-entry and splashdown. The launch is planned for no earlier than September 2025. Artemis 3 (2026) is arranged to be the primary manned lunar landing. The mission depends on a back mission to put a Starship Human Landing Framework (HLS) into a near-rectilinear corona circle (NRHO) of the Moon earlier to the dispatch of SLS/Orion. After Starship HLS comes to NRHO, SLS/Orion will send the Orion shuttle with a group of four to meet and dock with HLS. Two space travelers will transfer to HLS, which can slip to the lunar surface and spend around 6.5 days on the surface. The space travelers will perform at least two EVAs on the surface sometime recently the HLS rises to return them to a meet with Orion. Orion will return the four space explorers to the Earth. Dispatch is planned for no earlier than September 2026.

Artemis 4 (2028) is arranged to be the moment manned lunar landing mission. Orion and an updated Starship HLS will dock with the Lunar

Door station in NRHO before the landing. An earlier back mission will provide the primary two Portal modules to NRHO. The additional control of this mission's SLS Piece 1B will permit it to convey the I-HAB Door module for the association to the Door. Dispatch is planned for no earlier than September 2028. Artemis 5 (2029) is arranged to be the third manned lunar landing, which can provide four space travelers to the Door Space Station. The mission will convey the European Space Agency's ESPRIT refueling and communications module and Canadarm3, a Canadian-built mechanical arm framework for the Portal. Moreover conveyed will be NASA's Lunar Landscape Vehicle. Dispatch is scheduled for no earlier than September 2029. The mission will be the primary to utilize Blue Origin's Blue Moon lander to bring space travelers down to the Moon's surface.

Bolster missions incorporate mechanical landers, conveyance of Portal modules, Portal coordinations, conveyance of the HLS, and conveyance of components of the Moon base. Most of these missions are executed beneath NASA contracts to commercial suppliers. Beneath the Commercial Lunar Payload Administrations (CLPS) program, a few automated landers will provide logical rebellious and mechanical meanderers to the lunar surface after Artemis 1.

Extra CLPS missions are arranged all through the Artemis program to convey payloads to the Moon base. These incorporate living space modules and rovers in the back of manned missions. A Human Landing Framework (HLS) may be a shuttle that can pass on team individuals from NRHO to the lunar surface, bolster them on the surface, and return them to NRHO. Each manned landing needs one HLS, even though a few or all of the shuttle may be reusable.

Each HLS must be propelled from the Soil and conveyed to NRHO in one or more dispatches. The beginning commercial contract was granted to SpaceX for two Starship HLS missions, one uncrewed and one manned as part of Artemis 3. These two missions each require one HLS launch and numerous fuelling dispatches, all on SpaceX Starship launchers. Each Winged Serpent XL will stay connected to Portal for up to six months. The Dragon XLs will not return to Soil but will be arranged off, likely by deliberate crashes on the lunar surface.

6.1: Cultural Impact of Lunar Exploration

The lunar investigation has had a significant social effect on society, affecting different angles of human life, from science and innovation to craftsmanship, writing, and worldwide viewpoints. The lunar investigation has propelled headways in science and innovation. The challenges of space investigation have driven developments in areas such as mechanical technology, materials science, broadcast communications, and computer innovation, driving spin-off innovations that advantage different businesses on Earth.

The space race, including the Apollo missions to the Moon, got to be an image of national pride and ability amid the Cold War period. Accomplishments in space investigation were seen as pointers of a nation's logical and mechanical capabilities, contributing to a sense of national character.

Lunar Lander Quest

The lunar investigation has risen above national boundaries, cultivating a shared interest in space among individuals around the world. Worldwide collaboration in space missions, counting joint endeavors to investigate the Moon, reflects a collective human endeavor to investigate and get the universe.

Lunar investigation has started intrigued in science, innovation, designing, and arithmetic (STEM) areas. Famous minutes, such as the Apollo 11 Moon landing, have ended up as instructive points of reference, motivating eras of understudies to seek careers in space-related disciplines. The Moon and lunar investigation have become capable images in social symbolism. The Moon has been a repeating theme in craftsmanship, writing, and prevalent culture, speaking to riddles, motivation, and the human spirit's capacity for investigation. The visual effect of lunar investigation, captured through photos and film of Moon arrivals, has affected creative expression. Specialists have drawn motivation from the lunar scene, and the Moon has been included conspicuously in depictions, figures, and other shapes of creative creation.

The lunar investigation has ended up being a repeating subject in prevalent culture, showing in writing, movies, music, and TV. From science fiction stories about Moon colonies to documentaries describing chronic lunar missions, the Moon proceeds to capture the creative ability of groups of onlookers around the world. The prospect of a future lunar investigation, including the potential for space tourism, has captured the public's consideration. Private companies investigating lunar travel and missions have started to intrigue and engage individuals who dream of wandering past Earth's climate.

The "Earthrise " photo, taken amid the Apollo 8 mission, gave a special viewpoint of Soil rising over the lunar skyline. This picture has ended up an image of natural mindfulness and the delicacy of our planet, cultivating a sense of worldwide solidarity and obligation. The lunar investigation has fueled logical interest and a sense of pondering approximately the universe.

The ponder of the Moon's topography, composition, and history contributes to our broader understanding of the sun-oriented framework and past. The lunar investigation has not as it were extended our logical information but has moreover had an enduring effect on culture, affecting human imagination, creative ability, and our collective sense of character and investigation.

CONCLUSION

As we conclude our investigation of the Lunar Lander Journey and humanity's momentous travel to touch the Moon, we discover ourselves standing at the crossing point of logical accomplishment, mechanical development, and the persevering soul of investigation. The journey for lunar landers has taken us on a captivating journey, one that rises above borders, motivates eras and reshapes our understanding of the universe.

Our travel started with the Moon's allure—an old interest that saturated myths, fables, and the human creative ability. From the early lunar dreams established in social imagery to the logical interest that drove us to open the Moon's secrets, the chapters unfurled with stories of investigation, development, and the interest of the obscure.

The Space Race, a characterizing chapter in our lunar story, saw the strong competition between superpowers, coming full circle within the memorable Apollo missions. Apollo 11, with its notorious "monster jumps for mankind," speaks to the apex of human accomplishment, demonstrating that the incomprehensible was inside our get a handle on. The impressions cleared out on the lunar surface resounded over the unfathomability of space, symbolizing the unyielding human soul.

As we dug into lunar lander innovation, we saw the fastidious arranging, testing, and prototyping required to navigate the unforgiving lunar environment. Exact arrivals, independent frameworks, and cutting-edge impetus got to be the building pieces of the lunar investigation design, setting the arrangement for the following chapters in our enormous adventure.

Enter the Artemis program, a modern-day journey calling us back to the Moon. Named after the Greek goddess and sister to Apollo, Artemis carries the burn of investigation forward. With an emphasis on supportability, worldwide collaboration, and the guarantee of returning people to the lunar surface, Artemis is balanced to engrave unused chapters within the lunar logbook.

The social effect of lunar investigation is woven into the texture of our worldwide character. From the logical and mechanical motivation that fills advancement to the creative expression and imagery that penetrate our culture, the Moon has cleared out a permanent check on humanity's collective awareness.

As we mull over the conclusion of this Lunar Lander Journey, we discover ourselves on the brink of a modern time. The chapters however to be composed will unfurl with the Artemis program, private wanders, and the proceeded investigation of the Moon. The appeal of the Moon continues, drawing us ever closer to the universe and reminding us that our travel, as a species, is inflexibly connected to the firmament bodies that decorate the night sky.

Within the terrific embroidered artwork of human history, the Lunar Lander Journey stands as a confirmation of our intrinsic interest, our capacity for development, and our ceaseless want to reach for the stars. The Moon, once a far-off muse, has gotten to be a substantial waypoint in our investigation of the universe. As we conclude this chapter, we energetically expect the unfurling story of humanity's continuous journey to touch the Moon and the past.